W9-AMQ-195

Cover design by Allison Sundstrom/Sourcebooks
Internal design by Will Riley

Published by Sourcebooks eXplore, an imprint of Sourcebooks Kids
P.O. Box 4410, Naperville, Illinois 60567-4410
(630) 961-3900
sourcebookskids.com
First published as Red Kangaroo's Thousands Physics Whys: *Bathing in Waves: Wave Physics* in 2018 in China by China Children's Press and Publication Group.
Library of Congress Cataloging-in-Publication Data is on file with the publisher.
Source of Production: PrintPlus Limited, Shenzhen, Guangdong Province, China
Date of Production: February 2020
Run Number: 5017054
Printed and bound in China.
PP 10 9 8 7 6 5 4 3 2 1

Let's Ride a Wave!

Diving into the Science of Light and Sound Waves with Physics

#I Bestselling
Science Author for Kids
Chris Ferrie

What a perfect summer day to be at the beach! Red Kangaroo loves watching the waves. It's so relaxing!

"I wonder if the waves ever stop?" she thinks to herself. "I'll ask Dr. Chris!"

Dr. Chris is in his lab, setting up several things on a table to help answer her question.

"The waves never stop," he says. "Nature makes all types of waves in so many different sizes. All of these things make waves too."

“Really?” Red Kangaroo asks. “I guess I need to learn some wave physics!”

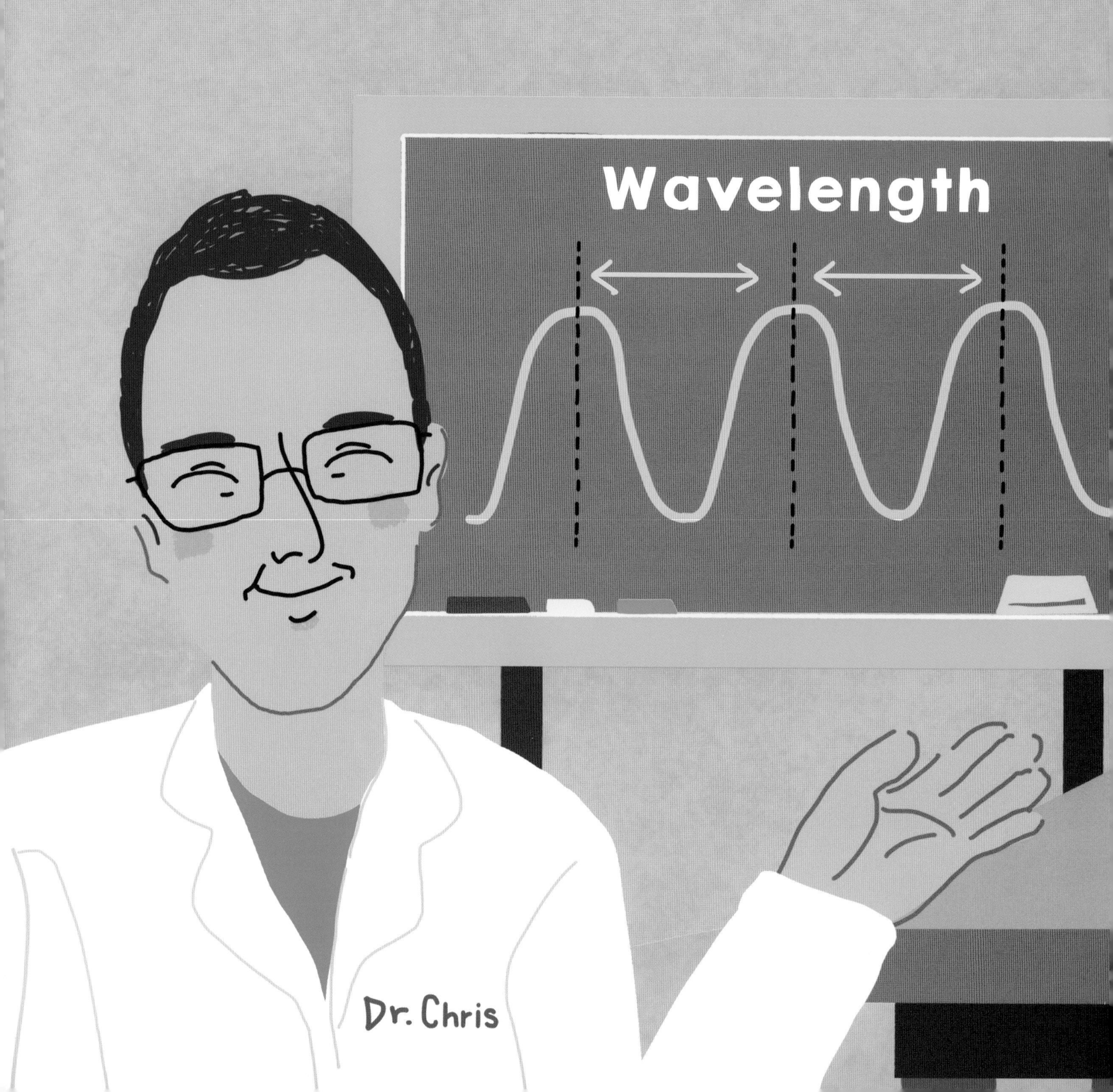
Wavelength
Dr. Chris

"First let's start with the definition of a wave," Dr. Chris says. "A **wave** is any repeating pattern. When we measure the space between each high point, that is called the **wavelength**."

"Oh, I get it!" Red Kangaroo exclaims. "So some waves will have short wavelengths and some have long ones?"

"That's right, Red Kangaroo!" Dr. Chris replies. "Let's go back to the beach to investigate!"

"Ocean waves are an example of waves you can easily see in nature," Dr. Chris explains. "The wavelength of these waves could be about as long as a football field. If we waited long enough, we would see that the whole ocean itself rises and falls twice each day."

"I know!" Red Kangaroo interrupts. "That's called the **tide**."

"That's right," says Dr. Chris. "The tide is a wave with a wavelength the size of the Earth!"

Dr. Chris

When they get back to the lab, Dr. Chris shines a laser on the wall.

"Is that light a wave too?" Red Kangaroo asks.

"Yes! All light is a type of wave," Dr. Chris replies. "And each color of light has a different wavelength."

"But why can't I see the wave?" Red Kangaroo wants to know.

"You *are* seeing a wave!" Dr. Chris says. "Light is a type of **electromagnetic** wave. Electromagnetic waves travel through space at such a fast speed that it's hard for us to recognize them as waves."

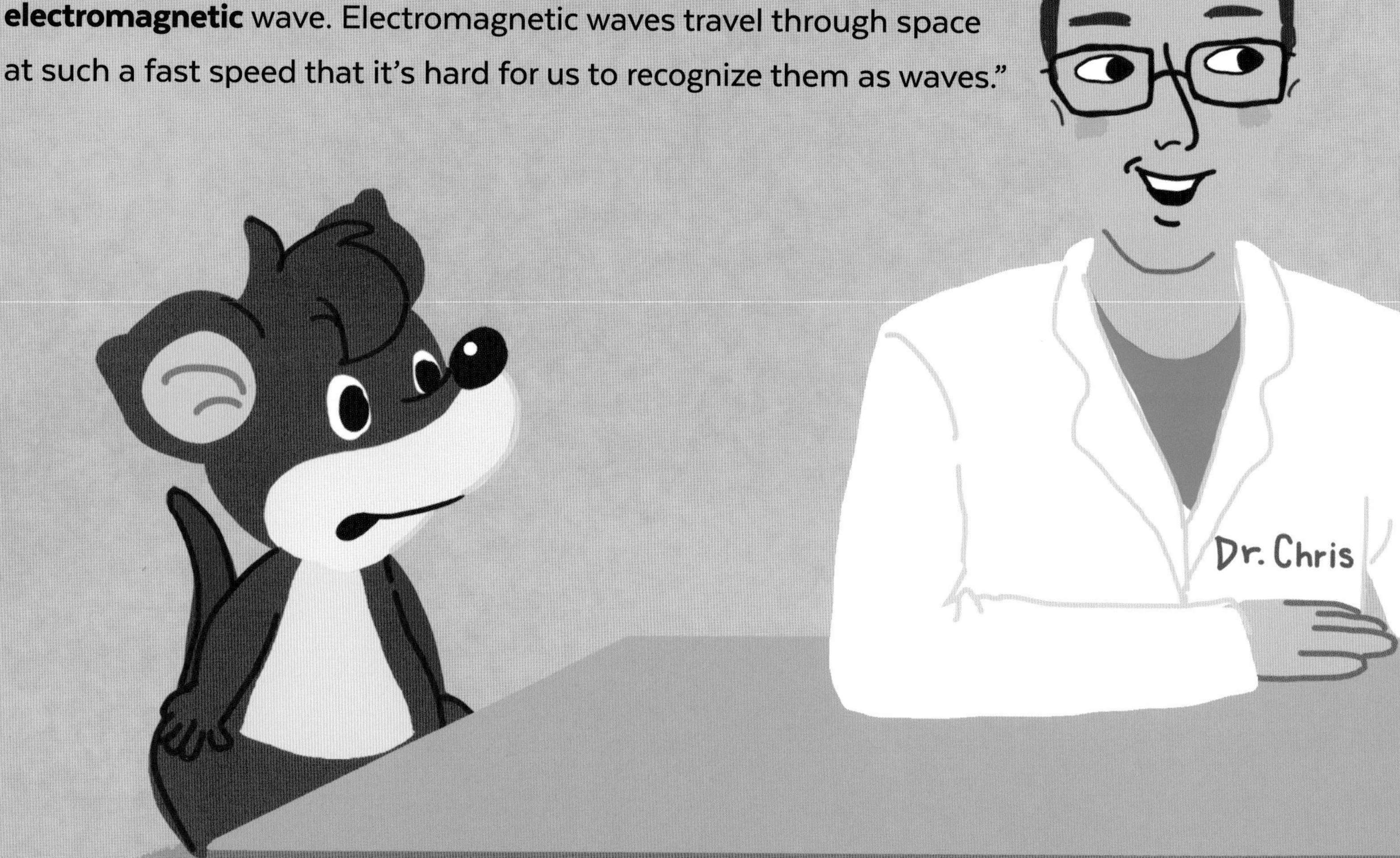

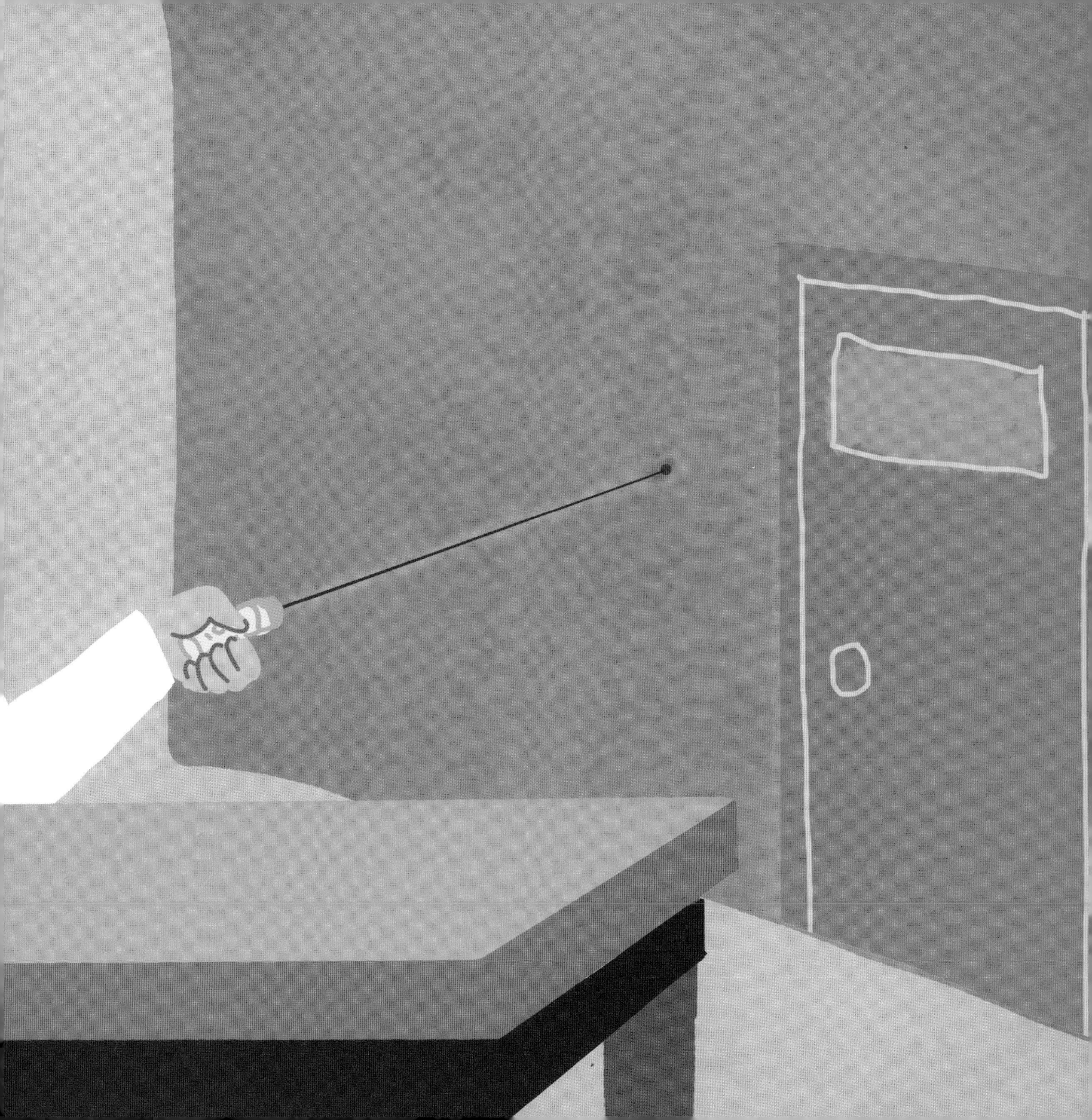

"Electromagnetic waves can have very different wavelengths," Dr. Chris continues. "Some of these waves you can see, like **visible light** and all the colors light helps us see. But there are also electromagnetic waves that you can't see at all."

"There are waves I can't see?" asks Red Kangaroo. "Can you teach me about them?"

Visible Light Spectrum

Electromagnetic Waves

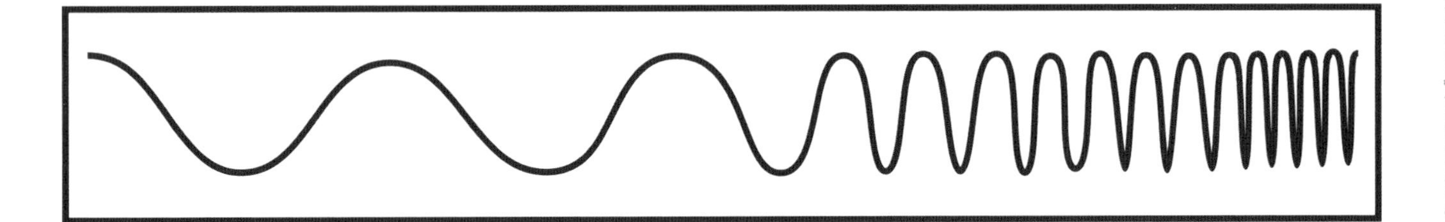

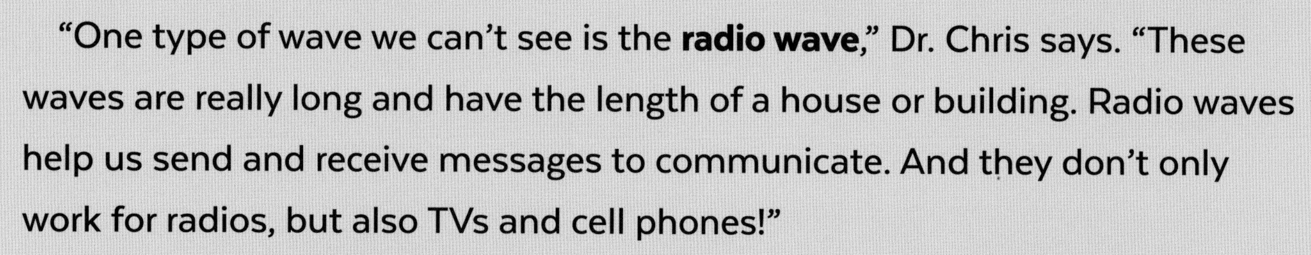

"One type of wave we can't see is the **radio wave**," Dr. Chris says. "These waves are really long and have the length of a house or building. Radio waves help us send and receive messages to communicate. And they don't only work for radios, but also TVs and cell phones!"

"Wow! So that's how I can get a call on my cell phone!" Red Kangaroo says.

"Another wavelength we can't see is the **microwave**," says Dr. Chris. "These waves are smaller than radio waves and are often used for cooking."

“They bounce around to make my hot chocolate warm! Yum!” says Red Kangaroo.

"Now let's talk about the waves that have wavelengths smaller than visible light," Dr. Chris says. "**Ultraviolet** light comes from the sun. We can also make ultraviolet light to grow plants indoors."

"I've heard of this wave before," Red Kangaroo says. "It's also called UV light. We have to wear sunscreen to protect our skin from UV rays!"

“And you probably have heard of X-rays. They have wavelengths even smaller than UV light,” Dr. Chris continues. “**X-ray** waves are useful to see inside things because they can travel right through your skin, but not your bones! You can find them at your doctor’s office, hospitals, and even airport security.”

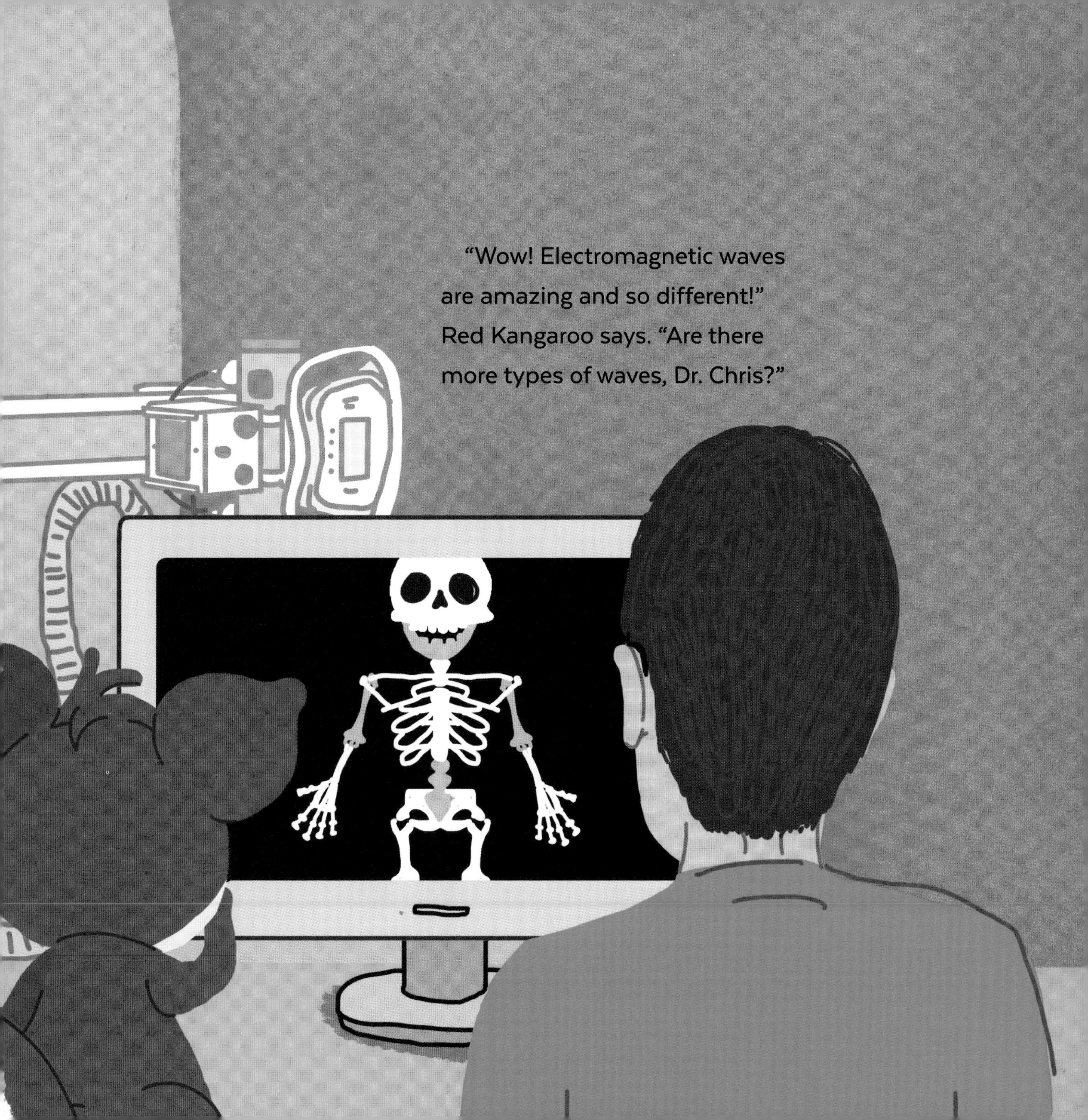

“Wow! Electromagnetic waves are amazing and so different!” Red Kangaroo says. “Are there more types of waves, Dr. Chris?”

“Yes, there are!” Dr. Chris replies. “**Sound** is made of waves that travel through air. You can’t see these waves either, but you can hear them!”

“But I thought light waves travel through the air too. What makes sound waves different?” Red Kangaroo asks.

Transverse Waves
Longitudinal Waves

"Great question!" Dr. Chris replies. "Water and light waves go up and down. We call them **transverse** waves. But sound waves go in and out and are called **longitudinal** waves."

Dr. Chris

"You can tell the wavelength of a sound wave based on how it sounds," says Dr. Chris. "Think about the sound of a drum, or noise from construction work. Those are low sounds, which means they have long wavelengths. Now think of the sound of a whistle or birds chirping. Those have high sounds and short wavelengths."

“Wow, Dr. Chris! I’ve learned so much about waves today!” Red Kangaroo says. “There really are waves everywhere!”

Dr. Chris
TAP!

Glossary

Electromagnetic wave
A wave of energy that travels through space and is made of electric and magnetic forces. Electromagnetic waves fall along a spectrum or range of wavelengths of different sizes.

Longitudinal wave
Waves that go in and out, squishing and stretching. Sound waves are longitudinal waves.

Microwave
A type of electromagnetic wave that can be used to cook food. It can also be used to communicate information and even help predict the weather.

Radio wave
A type of electromagnetic wave that is used to transfer information and data. Radio waves have long wavelengths and are used for things like radios, TVs, cell phones, satellites, and the internet.

Sound wave
A wave of vibrations in the air that we can hear. Sound can move through water and anything else!

Tide
The large, repeated rising and falling of the ocean level. This is a type of wave because the pattern repeats itself twice a day.

Transverse wave
Waves that go up and down. Light, heat, and water waves are examples of transverse waves.

Ultraviolet

A type of electromagnetic wave that comes from the Sun. We can't see ultraviolet (UV) light, but some insects like bees can.

Visible light wave

A type of electromagnetic wave that can be seen by the human eye. Every color we can see is because of a visible light wave.

Wavelength

The distance between one peak to the next as a wave pattern repeats itself.

Waves

Any pattern that repeats itself over time or as it moves through space.

X-ray

An electromagnetic wave with a very short wavelength that can go through things. This makes them useful for doctors to see inside our bodies and for security guards to look inside suitcases at the airport.

Show What You Know

1. Think about the definitions of wave and wavelength. Explain how they are different from each other.
2. Name two different types of electromagnetic waves.
3. Name the electromagnetic wave that can give you a sunburn.
4. Imagine you were a doctor. What type of wave would you use to see a patient's bones?
5. Explain the difference between a longitudinal wave and a transverse wave. What type of wave is sound? What type of wave occurs in water?

Answers on the last page.

Test It Out

Sounding it out!

Get together different objects from around your house. If you think you need special objects, guess again! You can use bottles or cups, your toys, or even doors and windows. Then find something you can use to gently tap on these objects like a wooden spoon, a pen, or even your knuckles!

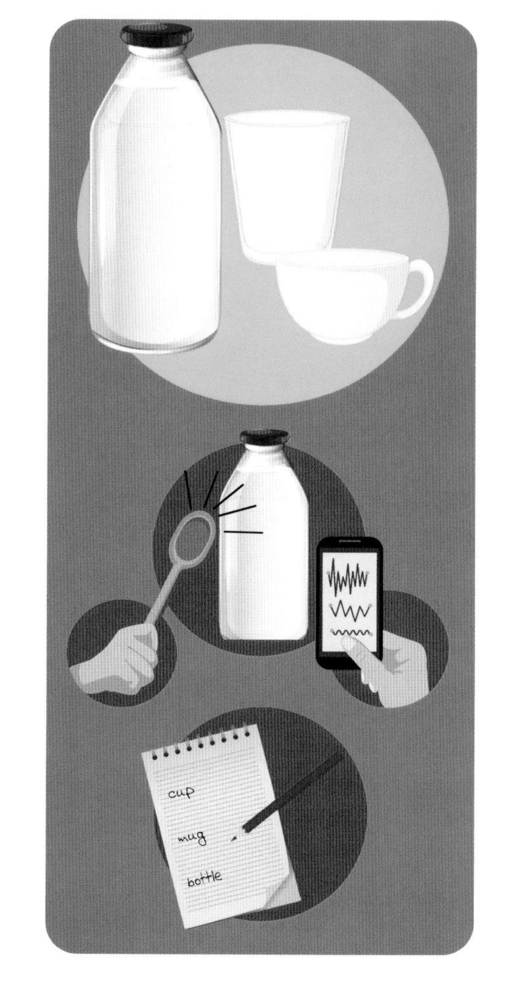

Think about what type of sound you think each object will make when you lightly tap on them. Make a prediction and try to list them in order from lowest to highest wavelength.

By listening carefully, you can test out your predictions by making some noise with each object! Record your results.

If you're having a hard time hearing the difference in sounds, download a science app that measures "audio spectrum." (Dr. Chris recommends trying the app Phyphox—ask an adult for help first!)

Compare your results to your predictions. How close were you in guessing the right order?

For a more controlled experiment, try this:
Gather a bunch of cups made of glass, metal, or porcelain that are the same size. Fill each cup with a different amount of water. They all will now produce different sounds when you tap on them, and you will see the waves in the water too!

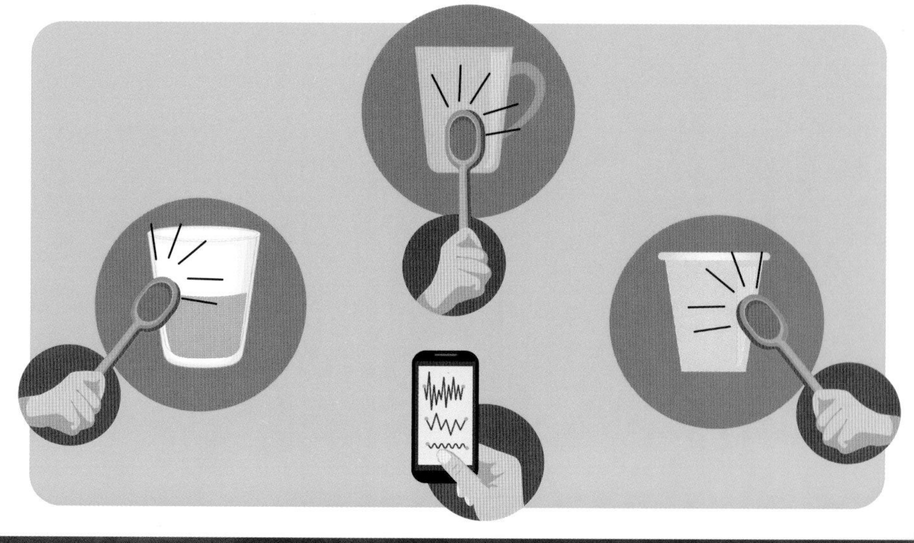

Every wave in one!

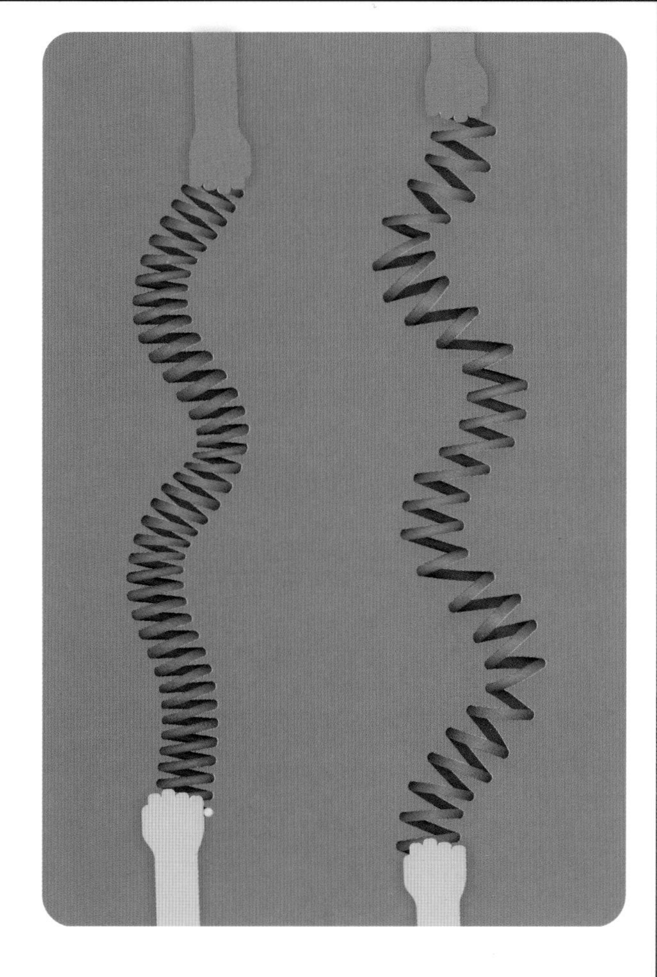

For this experiment, you will need a large spring toy (a metal one works best) and a friend. (You can also try this experiment with a rope, but it will be more challenging!)

Lay the spring toy on the ground and work together to stretch it out as far as you both want.

Have your friend hold their end of the spring toy still.

Shake, wiggle, and move your end of the spring toy to send waves! Try moving it in as many different ways as you can think of to create both transverse and longitudinal waves.

Try holding the spring toy out at different distances and continue making waves. Do you notice a change in the speed of the wave? Does the wave move more quickly when the two ends are closer together or farther apart?

Extend the fun: You can make it look like a wave is "standing still!" Think about a jump rope. When it's skipping, the wave doesn't "travel down" the rope. The crest of the wave is in the middle and never moves from there. That's called a "standing wave." If you have a jump rope nearby, you can try it. How many standing waves can you make? How does the number of standing waves depend on how fast you swing the rope?

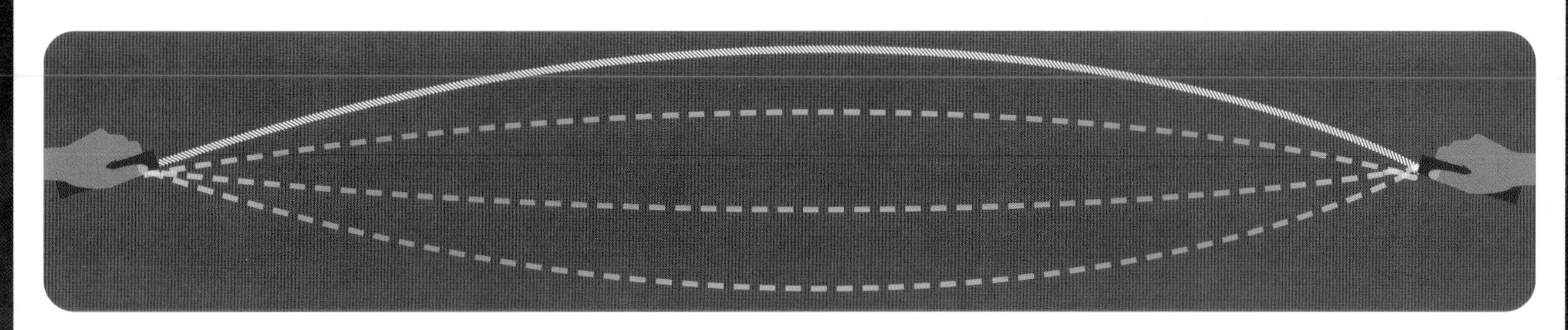

What to expect when you Test It Out

Sounding it out!

Different materials will make different sounds. Knocking on a door or tapping a shoe on the floor will make a lower sound than tapping on a window or clinking two glasses together.

Every wave in one!

The speed in the waves should be faster when you and your friend are farther apart. They will move slower when you are closer together. Swinging the spring toy side to side creates a transverse wave. Pulling and pushing the spring toy in and out makes a longitudinal wave. (Note that you can't make a longitudinal wave with a rope.)

Show What You Know answers

1. A wave is the repeating pattern, while a wavelength is the distance between each high point in the pattern
2. You read about radio waves, microwaves, visible light waves, ultraviolet waves, and X-rays.
3. Ultraviolet (UV)
4. X-ray
5. A longitudinal wave squishes and stretches. A transverse wave goes up and down. Sound is a longitudinal wave. Water makes transverse waves.